BUTTERFLIES

BUTTERFLIES
BEAUTIFUL FLYING INSECTS

Julianna Photopoulos

amber
BOOKS

Published by Amber Books Ltd
United House
North Road
London
N7 9DP
United Kingdom
www.amberbooks.co.uk
Instagram: amberbooksltd
Facebook: amberbooks
Twitter: @amberbooks
Pinterest: amberbooksltd

ISBN: 978-1-83886-160-5

Project Editor: Michael Spilling
Designer: Keren Harragan
Picture Research: Justin Willsdon

Printed in China

Contents

Introduction

Butterflies and moths share a common ancestor, but these short-lived insects have evolved into distinct species over time. Of course, one of them has a better reputation than the other. But moths can be equally stunning with diverse forms and bright, colourful wings – some day-time species even being confused with butterflies. While we often think of moths as night creatures, some only come out in broad daylight. And though bees are the best-known pollinators, butterflies do their

fair share during the day while moths work unnoticed, mostly in the dark. In their astonishing, transformative life journey from caterpillars to adults, butterflies and moths have adapted remarkably to both harsh and lush habitats – from camouflage and mimicry to migrating, feeding, roosting and mating – which has allowed them to thrive on every continent (except Antarctica) for at least 200 million years.

The world of both butterflies and moths will undoubtedly surprise and amaze you in many ways. But a large number of species are in peril, so as well as celebrating their uniqueness and beauty, we need the hidden power of these insects to continue.

ABOVE:
Monarch butterfly feeding on a bright purple bloom.

OPPOSITE:
Madagascan sunset moth, *Chrysiridia rhipheus*, feeding.

Butterfly Species and Anatomy

Unlike many other insects, butterflies are admired for their beauty. Upon seeing them, a sense of joy and lightness is awakened within us, with our thoughts wandering to rebirth or a new beginning. With about 18,000 species, each butterfly is unique in colour, shape and size. Butterflies are normally active during the day and every year we long for their brief spring and summertime appearances.

Butterflies have a distinctive body plan, with big showy wings, an agile body, a mid-body or thorax, and a small head with a coiled mouthpart called a proboscis. The butterfly shares these characteristics with its less appreciated, typically night-time cousin, the moth. Both are part of the order called the Lepidoptera, derived from the Greek for 'scale wings', alluding to the tiny dust-like scales across their wings and bodies that give them their colours and patterns. All members of the group have six legs and four wings: one large pair in front, with thick, strengthening 'veins', and a smaller set behind with fewer 'veins'. Wings are obviously used for flight, but can also help butterflies camouflage, warn or deceive predators, and attract or communicate with other individuals.

Butterfly anatomy, especially the prominent club-tipped antennae, wingspan and a habit of holding wings upright when resting, shows that butterflies belong to a single distinct superfamily called Papilionoidea. This group is generally considered to be broken down into six sub-groups, in which members share some physical features and behaviours.

OPPOSITE:

Common brimstone

A close-up of the head and mouthparts of the common brimstone, *Gonepteryx rhamni*, shows the long, curled, straw-like mouthpart called the proboscis. This uncoils when the butterfly sips liquids like nectar from flowers. *G. rhamni* is one of the longest-lived butterflies, living up to 13 months in Europe, North Africa and Asia. It is thought that the male's bright yellow wings inspired the name '*butter*-fly'.

LEFT AND ABOVE TOP:

Apollo

Named after the Greek god of many things including music, poetry and light, the species *Parnassius apollo*, or mountain Apollo, has a white hairy body and white wings with spots. Its red spots vary in size and shape, depending upon the subspecies, but often fade in the sun over time. At least 23 subspecies exist, but numbers in some parts of Europe are declining at an alarming rate.

ABOVE BOTTOM:

Arctic fritillary

The alpine species *Boloria chariclea* is a 'fritillary' butterfly. The Latin word for 'dice-box' or 'chequerboard' is *fritillus*, which lends itself to describe the chequered patterns on the wings, commonly black on orange. Most fritillaries look very similar so wing patterns must be closely examined to identify each species.

Banded orange heliconian
Also known as the orange tiger, this species (*Dryadula phaetusa*) takes its name from its bright orange and black-striped wings. The bright colour warns predators to stay away, but the butterfly likely mimics more poisonous and harmful species. *D. phaetusa* belongs to the largest butterfly family, Nymphalidae, known as brush-footed or four-footed because their front legs are much smaller and are not used to stand or walk, and often have brush-like hairs on their feet.

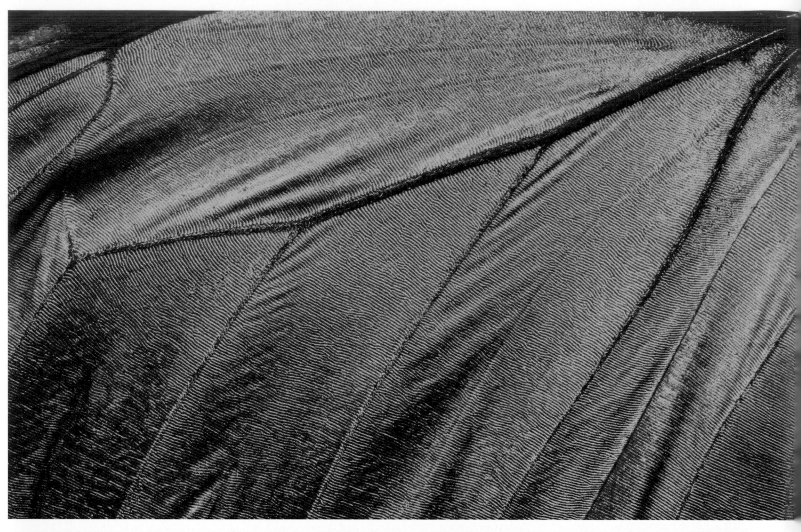

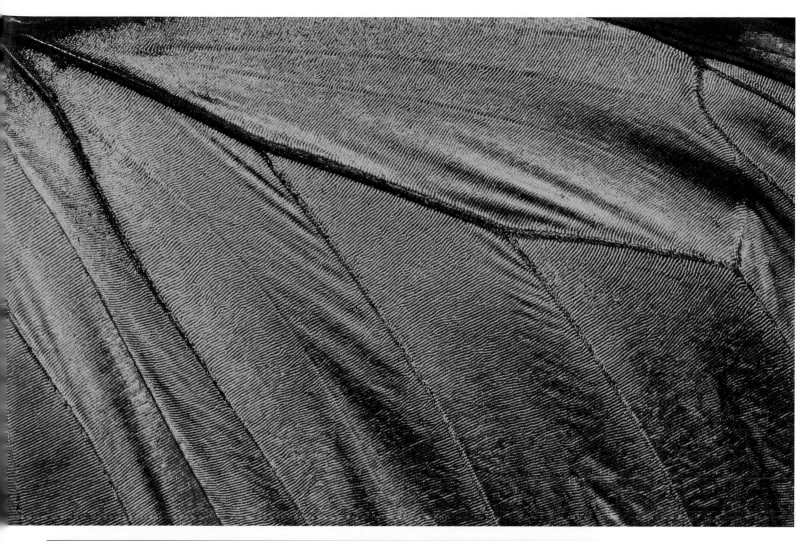

ALL PHOTOGRAPHS:
Blue morpho

As its name suggests, the tropical species *Morpho peleides* has iridescent blue wings with black edges. This shimmering blue colour is not a chemical pigment; it is produced by the way the light reflects off the millions of tiny scales on its wings. However, the wings underneath are a dull brown with eyespots, allowing blue morphos to blend in with their surroundings when resting with closed wings. They also escape predators with their flashing contrasting wings while flying. With wing spans of 13–20cm (5–8in), *M. peleides* is one of the world's largest butterflies.

Blue tiger
The species *Tirumala limniace* has a unique bluish-white spot wing pattern, differentiating it from other tiger species and their mimics. Its head and thorax is brownish black with white dots. Blue tiger caterpillars feed on toxic milkweed plants, making them, and adult butterflies, poisonous to predators.

ABOVE TOP:

Chimaera birdwing

This New Guinean species, *Ornithoptera chimaera*, lives on the New Guinea Highlands mountain range. The *Chimaera* was a Greek mythological creature made up of three animals and has come to describe something unbelievable or awe-inspiring. Males and females differ in size and colour – this colourful male is smaller than the dark-brown, white-spotted female.

ABOVE BOTTOM AND RIGHT:

Common blue

Living up to its name, *Polyommatus icarus* is one of Europe's most common blue butterflies. While the male has blue upperside wings with a black-dark-brown border and white fringe, the female is brown with orange-red spots and some blue that varies significantly in individuals. As members of the Lycaenidae, or gossamer-winged, family, the second-largest family with over 6,000 species, common blues have small wingspans of less than 5cm (2in), and often have white-ringed antennae and eyes.

ABOVE AND OPPOSITE TOP:

Cracker

Members of the *Hamadryas* genus
are known as cracker butterflies.
However, only the males live up to
this name by making an unusual
cracking noise with their wings to
mate or defend their territories.
Their cryptic colouration helps
them blend in with tree trunks or
boulders when resting with their
wings open.

OPPOSITE BOTTOM:

Common buckeye

The species *Junonia coenia* has
conspicuous large eyespots on
its wings, resembling targets. It
goes through many generations
each year and the colours of its
underside hindwings depend on
the season: those emerging in
the summer have beige or brown
with more distinguished patterns,
whereas winter butterflies are a
vivid reddish-brown.

ALL PHOTOGRAPHS:
Dead leaf

Named for its dry, brown leaf appearance when its wings are closed, this species, *Kallima inachus*, cleverly tricks birds, ants, spiders and wasps to avoid being eaten. But when open, its wings reveal a dazzling colour pattern with oranges, browns and blues. The dead leaf, also known as the orange oakleaf, changes colour and size to blend in with its environment, depending on the season. During the wet season, its wings are a darker brown, while the coloured side is a richer royal blue with thick, bright bands of orange. Its dry-season form is more muted.

ABOVE AND OVERLEAF:

Eastern comma

This North American species, *Polygonia comma*, is a common visitor to rotting fruit. Its closed wings, which mimic a fallen leaf, have a silvery mark that resembles a comma. The other side of its wings are orange and dark brown or black, with colours varying between winter and summer. Species belonging to the *Polygonia* genus are known as anglewing butterflies because of their irregular wing edges.

OPPOSITE:

Anna's 88

As its name suggests, *Diaethria anna* has a black outlined '88' on the underside of its red, white and black-banded wings. Sometimes the number looks like an '89' or '98'. Its upperside is dark brown with a wide, metallic green band on the forewings. *D. anna* is native to wet tropical forests in Central America, but is occasionally found in southern Texas.

Emerald swallowtail
Also known as the banded
peacock, or emerald peacock,
the blazing emerald-green band
that runs across each dark-green-
black wing gives the *Papilio
palinurus* species its name. Its
wings underneath are black with
orange-and-white spots along the
edges. *P. palinurus* belongs to the
swallowtail family, or Papilionidae,
with at least 550 species, known
for their often long, extended tails
at the bottom of their wings.

PREVIOUS PAGE:
Forest giant owl
Named for its huge eyespots that resemble an owl's eyes, the forest giant species, *Caligo eurilochus,* has a wingspan up to 17cm (6.7in). The eye pattern on its wings is thought to scare predators away. *C. eurilochus* can be found all year round but is only active at dawn and dusk.

ABOVE:
Hecale longwing
The subspecies *Heliconius hecale zuleika* is known as the tiger, the golden or the Hecale longwing butterfly. It lives across Southeast Mexico all the way to Panama. Many members of the *Heliconius* genus get their amino acids – needed to produce eggs – by feeding on pollen as adults. This allows them to live several months longer.

RIGHT:
Glasswing
The migratory *Greta oto*, or glasswing butterfly, has transparent wings, enabling it to hide in plain sight. These wings are sparse, spindly scales covering an anti-reflective waxy-coated membrane. Found across Central and South America, these butterflies are popularly known in Spanish-speaking regions as *espejitos*, or 'little mirrors'.

Kaiser-i-Hind
With its glimmering greens, bright yellows and fine blacks, this rare swallowtail species, Kaiser-i-Hind, or *Teinopalpus imperialis*, is nicknamed the 'Emperor of India'. It is protected by Indian and Nepalese law but still sought after by butterfly collectors.

LEFT AND ABOVE TOP:

Malachite

This species, *Siproeta stelenes*, is named after the green mineral, malachite. Its wings are dark brown-black and translucent green or yellow-green, while the underside is a light brown and olive green. It prefers feeding on rotting fruit, and sometimes on bird droppings and nectar. At night, malachite butterflies roost together in low shrubs.

ABOVE BOTTOM:

Map

The *Araschnia levana* butterfly of the summer brood perched on a stinging nettle – the caterpillar's food plant. These charming butterflies are unique in that they have two seasonal forms: the spring brood *levana* and summer *prorsa*. When the wings are open, the summer brood is black with white marks, whereas the spring brood is orange and looks like small fritillary butterflies.

ABOVE:

Monarch

The distinctive orange colour of the migratory monarch is thought to have given rise to its name, honouring King William III of England whose secondary title was 'Prince of Orange'. Its wing pattern – orange laced with black lines and bordered with white-dotted black bands – warns predators that they are poisonous. This species, *Danaus plexippus*, is similar to two other monarch species: the *D. erippus* (southern monarch) and *D. cleophile* (Jamaican monarch).

OPPOSITE TOP AND BOTTOM, AND OVERLEAF:

Painted lady

This migratory species, *Vanessa cardui*, or painted lady, is also referred to as the thistle butterfly because its caterpillars thrive on these plants. This well-known butterfly has orange-brown wings with white spots on black patches, whereas underneath they have a grey, black and brown marbled pattern with four black spots. Painted ladies can detect ultraviolet, blue and green light, but not red like the monarch can.

Peacock

The European peacock, or *Aglais io,* was previously the only member of the now defunct genus *Inachis*. Io was a priestess of the ancient Greek goddess Hera, who was desired by her husband Zeus and metamorphosed into a white cow to avoid Hera's suspicions. Hera honoured her giant watchman Argus, who guarded Io and was killed under Zeus' orders, by putting his 100 eyes into the peacock's tail feathers, which also resemble this butterfly's eyespots.

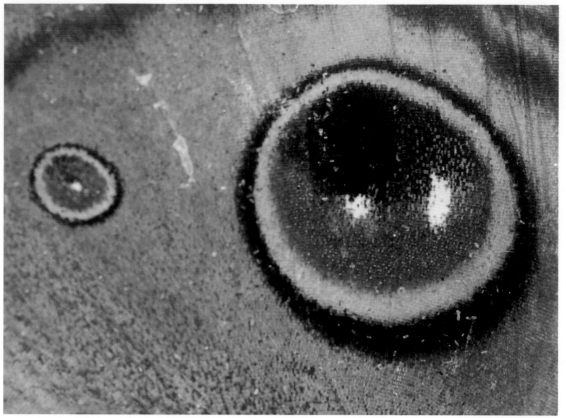

ABOVE TOP AND RIGHT:

Peacock pansy

Junonia almana, or the peacock pansy, lives in India and Southeast Asia, spreading to China and Japan. Like the common buckeye, the peacock pansy has two distinct pattern forms during the wet and dry seasons. Its wings are bright orange with notable eyespots on the hindwings, becoming more profound in the wet season, while the undersides are paler but lack markings and eyespots in the dry season.

ABOVE BOTTOM:

Piedmont ringlet

This alpine species, *Erebia meolans*, is native to Europe. It prefers rocky areas of mountains, often near forests, and is commonly seen flying at altitudes ranging from 900–1,800m (3,000–6,000ft). The Piedmont ringlet is easily identified by its dark wings with black-ringed white spots on the outer edges of red bands.

Purple emperor

This majestic male *Apatura iris* is an elusive species. It has iridescent purple wings, often appearing black from certain angles when the light is not being refracted from its wing scales. Females do not possess this purple sheen and instead are a dark brown. They both have a small orange ring on their hindwings, and white bands and spots. Their underwings are light brown, with the female having an orange-ringed eyespot on its forewings.

Question mark
The North American species *Polygonia interrogationis* has orange wings with dark spots and a powdery white border. However, a silver question mark appears when it closes its wings, giving it its common name. Sometimes the dot is missing and the dark form can be confused with that of the eastern comma's, but the two differ from each other by their size and the shape of the comma.

ALL PHOTOGRAPHS:
Rajah Brooke's birdwing
The *Trogonoptera brookiana* species is a protected birdwing butterfly named after Sir James Brooke, the 19th-century White Rajah of Sarawak. Birdwing butterflies are known for their elegant bird-like flight, great size and colourful males. These narrow-winged, electric-green and black male butterflies have a bright red head and their bodies are black with red markings. The larger females have the same patterns but are brown rather than black, and the green is replaced by a dull yellow.

Red admiral
This widespread species, *Vanessa atalanta*, lives across temperate northern Africa, Europe, Asia, North and Central America, and the Caribbean. It can be recognised by the broad red stripes and white-spotted tips on its black wings. Red admiral females only mate with males who hold a territory, and males guard their territories by chasing off intruders – often 10–15 times per hour.

Richmond birdwing
The Australian native species
Ornithoptera richmondia is known
for being a strong flyer, commonly
active in the early morning and
near dusk. Unlike the female,
this male Richmond birdwing
has distinctive iridescent green
markings. However both have
yellow bodies with a red
thorax – colours warning
predators of toxicity.

ALL PHOTOGRAPHS:
Silver-studded blue
Similar to the common blue, the male *Plebejus argus* species is blue with a dark border while the females are brown with orange-red spots. However, the silver-studded blue males usually have thicker black borders and additional metallic silver-greenish flecks under their wings. *P. argus* eggs and caterpillars are looked after by ants and, in return, the ants feed on sugar-rich honeydew secreted by the caterpillars.

Silver-washed fritillary
This species, *Argynnis paphia*, takes its name from the silver streaks under its wings. When opened, its wings are a vivid orange with black spots. The brighter orange males have four unique black veins on their forewings that bear scent scales to attract females. The females lay their eggs in tree bark cracks, close to their caterpillars' food – wood violets. When the eggs hatch, the caterpillars eat the empty eggshells and go into hibernation until spring.

ALL PHOTOGRAPHS:
Spanish festoon
The staggering species *Zerynthia rumina* lives in northern Africa, through the Iberian Peninsula, reaching southern France. In France, often confused with its close relative, *Z. polyxenia,* or the southern festoon, the Spanish festoon has striking red patches on its forewings and lacks its cousin's blues on its hindwings.

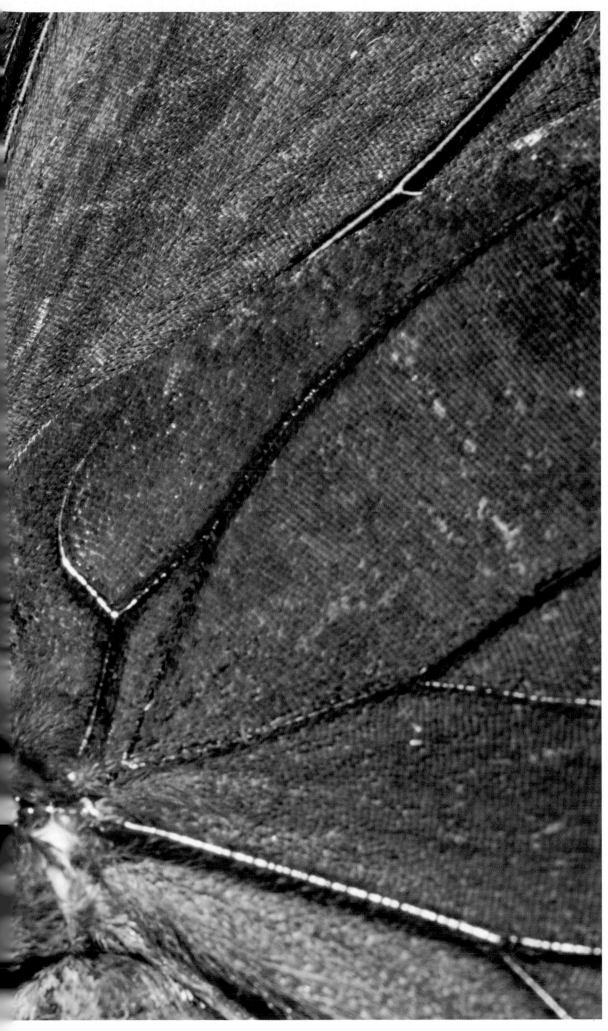

Ulysses
The swallowtail species *Papilio ulysses*, named after the Greek hero from the *Odyssey*, is also known as the blue emperor, or mountain blue, for its black-bordered iridescent electric blue wings. Its wings underneath are a dim black and brown for camouflage, but females have some extra blue crescents on the top tips – males are attracted to the blue and occasionally mistake other blues for females!

LEFT:

Old World swallowtail

The species, *Papilio machaon*, is brightly coloured and has 37 subspecies. Despite the term 'swallowtail' being the common name of all members of the Papilionidae family, the Old World swallowtail was the first to get the name.

ABOVE TOP AND BOTTOM:

Viceroy

Often confused with the monarch, the species *Limenitis archippus* is smaller, has an additional black line running across the veins on both hindwings and has a single – rather than double – row of white dots in the black marginal band. Similarly to the monarch, its wing pattern warns predators that it is foul-tasting and poisonous.

Moth Species and Anatomy

Moths are famous for being dull, drab pests, attracted to artificial light at night, and eating our clothing. But these insects will surprise you. Not only are they important pollinators, they also come in all shapes, colours and sizes, and are among the most diverse and successful creatures.

Moths have the same distinctive body plan as their more highly-regarded day-flying cousins, the butterflies: an agile body, a mid-body with six legs and four large wings, and a small, busy head with a coiled tube-like mouthpart. Nevertheless, there are about 160,000 moth species which constitute the biggest proportion – over 90 percent – of the order Lepidoptera, which they share with butterflies. In this group, insects have tiny coloured scales covering their wings and bodies.

Generally, the most reliable way of distinguishing moths from butterflies is by their antennae. Moths usually have thread-like or feathery antennae, while the antennae of butterflies are club-tipped. Unlike butterflies, moths typically fly during the night to slurp up nectar from flowers. While resting, their wings are folded horizontally flat over the body, with forewings and hindwings overlapping. However, there are many day-flying, brightly coloured moth species too, which can often be mistaken for butterflies. In fact, one of the most beautiful insects in the world is a day-flying moth from Madagascar. From colourful to cryptic, moths can be mouthpart-less, migratory, and be as small as a pinhead or as large as an adult's hand. Come, be dazzled by the world of moths!

OPPOSITE:
Atlas moth
This moth species, *Attacus atlas*, is one of the largest insects in the world with a wingspan extending up to 27cm (10.6in) across. Once it emerges from the cocoon, it does not feed at all – its sole purpose is to find a mate at night, within its one or two weeks of life. Atlas moths are cultivated in some areas of Southeast Asia for their strong, brown, woolly silk called *fagara*.

RIGHT AND OVERLEAF:
Atlas moth
Native to the tropical and
subtropical forests of Southeast
Asia, the atlas moth shares its
name with the Titan god of
Greek mythology. Its name in
Cantonese also means 'snake's
head moth' which refers to its
forewings looking like the head
of a snake, scaring off predators.
This large species belongs to the
Saturniidae family that consists
of about 2,300 species. Saturniids
have large, often colourful,
overlapping wings, feathery
antennae and tiny proboscises
that do not work for feeding.

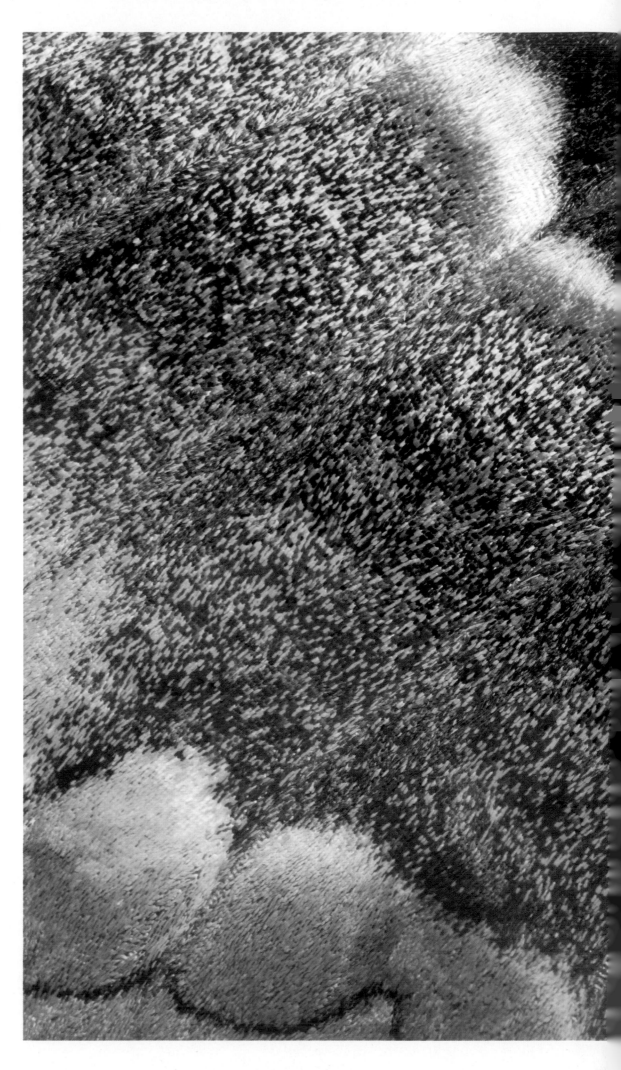

ALL PHOTOGRAPHS:

Cecropia moth

This species, *Hyalophora cecropia*, is the largest moth in North America with a wingspan of 12.7–17.8cm (5–7in). A member of the Saturniidae family, its mouthparts are not fully formed so it must take in enough food as a caterpillar to live for about two weeks. Male cecropia moths have an extraordinary sense of smell with their feathery antennae, detecting females' natural chemicals called pheromones over 1.6km (1 mile) away. However, bolas spiders can mimic these pheromones so moths can end up becoming dinner!

THIS PAGE AND OPPOSITE:
Cinnabar moth
Hundreds of black-and-white striped caterpillars can be found feasting on poisonous ragwort during the summer. By eating enough of this plant, cinnabar moth caterpillars become toxic themselves and their colourful stripes warn predators to stay away. If food becomes scarce, they may also end up eating their siblings. These caterpillars transform into the stunning, foul-tasting black-and-red species *Tyria jacobaeae*, or cinnabar moth, which flies during the day and night and is named after the red mineral cinnabar once used by painters.

Comet moth

Living for about a week in rainforests, the brightly-coloured Madagascan moon moths, or *Argema mittrei*, are often referred to as comet moths because of their long tails. The caterpillar, along with most members of the Saturniidae family, spins a silk cocoon. This loosely-fitted silvery cocoon is shaped like a sac and has many holes, thought to prevent pupae from drowning when it rains.

Blue tiger moth
This caterpillar, from India, belongs to the colourful moth genus *Dysphania,* of the Geometridae family. The family name is derived from the Greek for 'earth-measurer', alluding to the way the caterpillars, or inchworms, move. The inchworm will transform into the day-flying *Dysphania percota*, or blue tiger moth.

Elephant hawk-moth

The goldish-olive-pink elephant hawk-moth, or *Deilephila elpenor*, is an important pollinator of many habitats, including grasslands, across Europe and Asia. These moths can see colour in low light allowing them to hover over flowers and drink nectar at night. Elephant hawk-moth caterpillars look like elephant trunks, which gives these moths their name.

Emperor moth
A fluffy colourful insect that inhabits heaths, grasslands and sand dunes, *Saturnia pavonia*, or small emperor moth, lives across Eurasia from northern Africa and the foothills of the Himalayas to Japan. It is the only member of the saturniids that lives in the United Kingdom, where it is known as the emperor moth. Males fly during the day searching for female scents with their feather-like antennae, while females lay low and only come out at night.

ABOVE AND OPPOSITE TOP:
Galium sphinx
Also known as the bedstraw hawk-moth, the species *Hyles gallii* is found in temperate areas of the northern hemisphere. From dusk until dawn, this moth hovers in mid-air while feeding on nectar from flowers. Most sphinx (or hawk-moth) caterpillars have a curved 'horn' on their tail ends. The galium sphinx is named after the *Galium* genus of plants, which it feeds on as a caterpillar.

OPPOSITE BOTTOM:
Garden tiger moth
This furry caterpillar, commonly called 'woolly bear', belongs to the garden tiger moth. Covered in long, thick black-and-ginger hairs that cause irritation to those who touch them, they can safely munch on stinging nettles, dock leaves, foxglove and many garden plants.

Garden tiger moth
Up close, the mandibles of a
garden tiger moth caterpillar
look strong. These are used to
slice their way through often
poisonous leaves – the toxins are
forever stored in the insect's body,
protecting it from predators. After
transforming into the colourful
moth, *Arctia caja*, the mandibles
completely vanish. The moth's
bright colours warn predators
it is poisonous and foul-tasting.
A. caja is found across gardens,
meadows and woodland habitats
in Europe, Asia, the United States
and Canada.

ALL PHOTOGRAPHS:
Giant leopard moth
This North American species, *Hypercompe scribonia*, is white with hollow and solid black or iridescent blue spots, a colourful body and black-and-white-banded legs. It has been known by many species names under the genus *Ecpantheria*. Giant leopard moths are found flying at night in fields, meadows and forest edges.

Hummingbird hawk-moth
Like its avian namesake, the
hummingbird hawk-moth (or
Macroglossum stellatarum), from
temperate Europe and Asia,
makes a humming noise while
hovering over tube-shaped
blooms and slurps up nectar with
its long proboscis from dawn
to dusk. Hummingbird hawk-
moths are not to be confused with
members of the North American
genus *Hemaris* – hummingbird,
or bee, moths.

Io moth

The colourful North American species *Automeris io*, or Io moth, belongs to the short-lived Saturniidae family, which are unable to feed. It is known for its distinct eyespots on the back wings, showing them off to scare predators away. As a caterpillar, the Io moth has poisonous stinging spines.

OPPOSITE AND ABOVE TOP:

Japanese silk moth

This saturniid moth, *Antheraea yamamai*, lives in east Asia. In Japan, it has been cultivated over 1,000 years for its strong, elastic and rare white silk. However, it was introduced to Sri Lanka, northern India and Europe for producing tussar silk, and can now be found in Austria, Italy and the Balkans.

ABOVE BOTTOM:

Lime hawk-moth

The species *Mimas tiliae* is found throughout Europe, Asia and northern Africa. A band of green across its wings helps the moth hide in its woodland habitat. Male lime hawk-moths are usually smaller and have more colourful markings than the females.

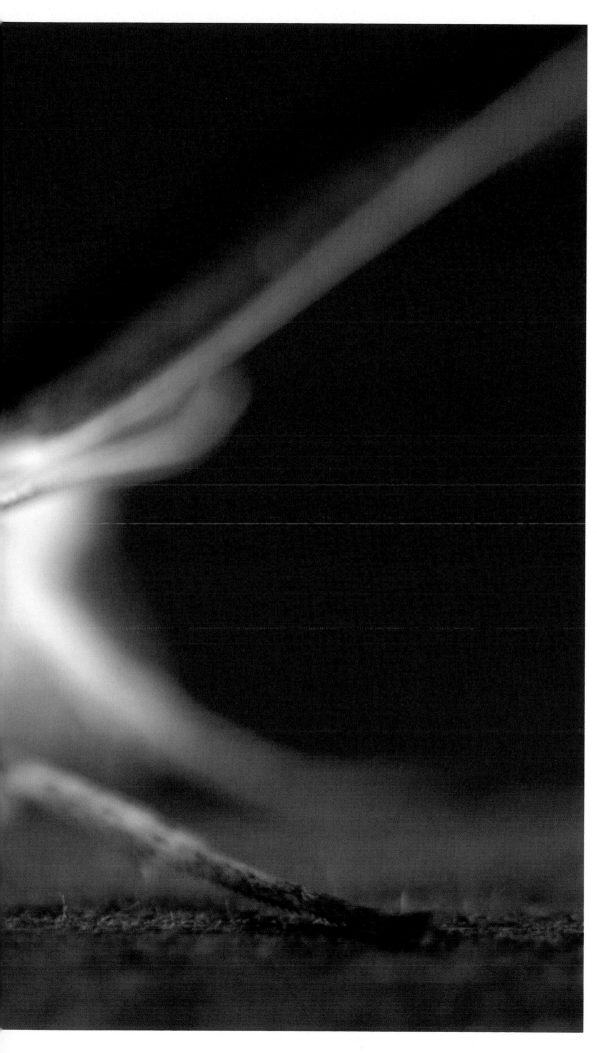

Hawk moths
Members of the Sphingidae family, with about 1,450 species, are known as the sphinx, or hawk moths. These moths have narrow wings and body forms that make them agile and fast flyers. Some species hover over flowers, resembling hummingbirds. Many of their caterpillars are known as 'hornworms', alluding to an unharmful 'horn' at the end of their bodies. Others have eye-marks that are accentuated when alarmed, giving the caterpillars a snake-like appearance.

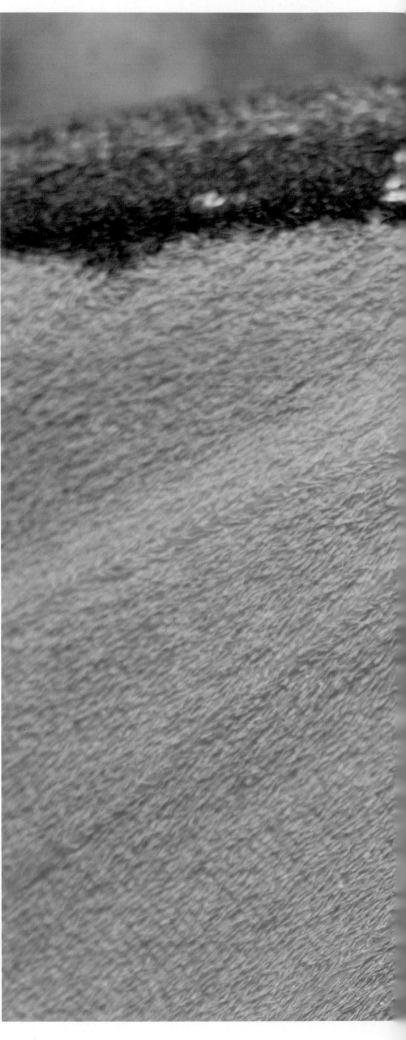

ALL PHOTOGRAPHS:
Luna moth
The green species *Actias luna*, from eastern North America, is named after the Roman moon goddess for its moon-like eyespots. While flying at night, luna moths spin their long tails in circles to confuse predatory bats and avoid being eaten. Like other saturniids, these moths lack mouthparts and a digestive system, living only for about a week.

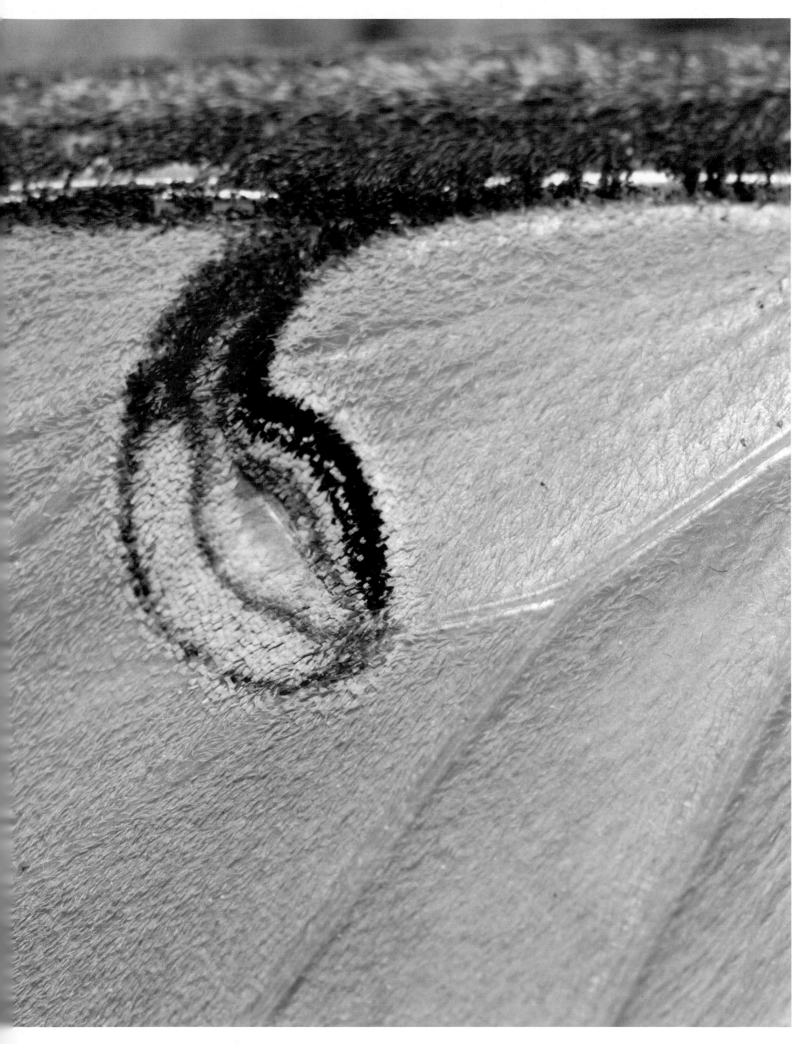

Owlet moths
Members of the Noctuidae family are known as owlet moths, with most species being active at night. Their caterpillars are called 'cutworms' or 'armyworms' because some of them gather together and destroy crops, orchards and gardens across every continent except Antarctica.

Oleander hawk-moth
The species *Daphnis nerii* grows
from a green hornworm to an
army green moth – another name
for the oleander hawk-moth. It
lives in Africa and Asia, but as
a migratory species it flies to
eastern and southern parts of
Europe during the summer. The
D. nerii caterpillar munches on
toxic oleander plants, while as
an adult it hovers over flowers to
slurp nectar after sunset.

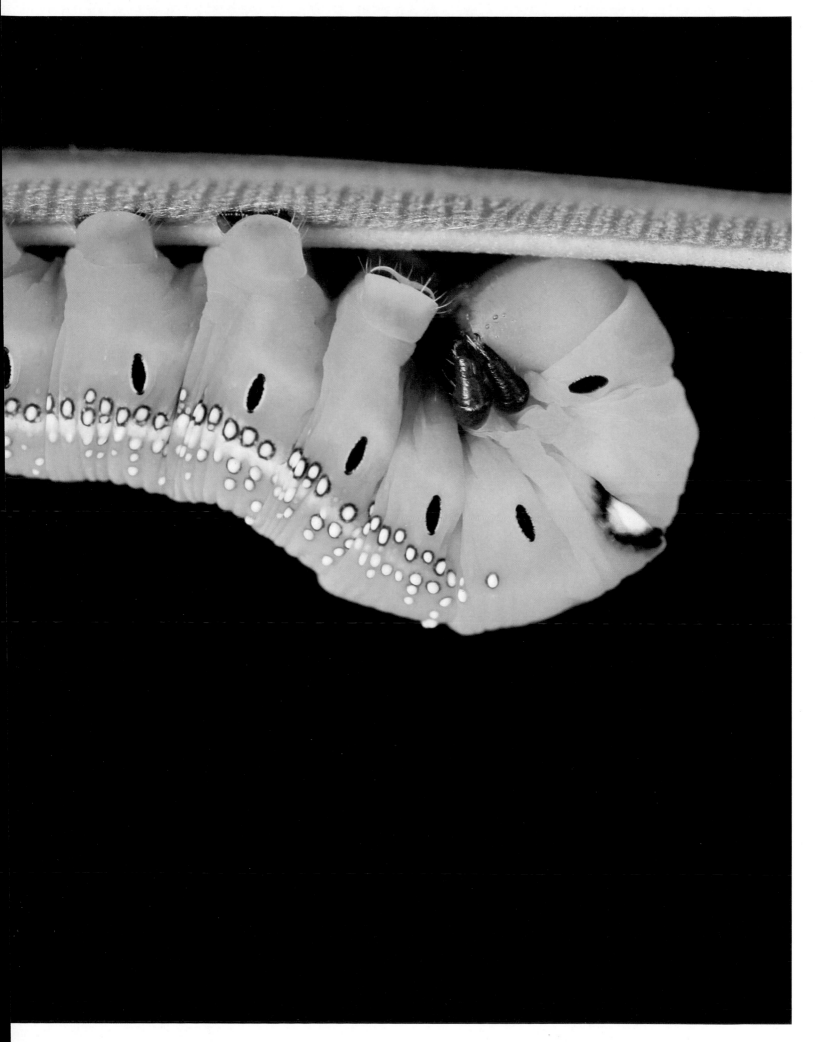

ALL PHOTOGRAPHS:
Pellucid hawk-moth
Cephonodes hylas is known by many names: coffee bee
hawk-moth, coffee clearwing and pellucid hawk-moth.
Its transparent wings and stout body make it look like a
bumblebee, but it hovers and sucks nectar from flowers like
other hawk-moths. *C. hylas* lives in many habitats across
Africa, India, Southeast Asia and Australia, devouring coffee
and pomegranate plants as a caterpillar.

Rosy maple moth
This yellow-and-pink North American species, *Dryocampa rubicunda*, is the smallest of the giant silk moths from the Saturniidae family. Its wingspan is 3.2–5cm (1.3–2in), with males being relatively smaller. Rosy maple moths live in or near woodlands and parks where maple trees grow.

LEFT:
Texas wasp moth

The species *Horama panthalon*, which lives from South America to the southern United States, mimics a stinging, paper wasp.

ABOVE TOP:
Uropyia meticulodina

A master of camouflage, this leaf-like moth lives in some parts of eastern Asia. Although its wings look curled up, it is an optical illusion!

ABOVE BOTTOM AND OVERLEAF:
White-lined sphinx

As its name suggests, the white-lined sphinx, or *Hyles lineata*, has white lines covering its wings' veins. A saturniid, the moth hovers over flowers like a hummingbird and sips nectar from dusk until dawn (and sometimes during the day). It can be found across most of North America – from Central America to southern Canada – the West Indies, Eurasia and Africa. *H. lineata* caterpillars have distinct colours and patterns but all have a harmless orange 'horn' on their backsides.

Habitats

Butterflies and moths can be found anywhere in the world, except Antarctica – from the tropics near the equator to northern regions above the Arctic Circle, and from sea level to high mountain tops. They live in every habitat on land, from tropical forests to deserts, grasslands, wetlands and alpine areas. Each species of butterfly and moth has its own region and habitat, as we'll see. Some have specific requirements and cannot live anywhere else, while others can adapt to new habitats. Many migrate short distances while some travel thousands of miles to their winter homes, then head back again in the spring.

A habitat has to meet all the environmental conditions that creatures need to survive. For butterflies and moths, that means finding and taking in food and shelter, selecting a mate and successfully procreating. It must also provide exact requirements for all their life stages: egg, larva, pupa and adult. The differences in behaviours and wing pattern colours illustrate how distinct species have adapted to their environments. Some blend in with flowers, leaves or bushes, while some mimic other species. Females and males can also differ from one another, making colours important for sexual attraction.

OPPOSITE:
Oileus giant owl
A cryptic butterfly, *Caligo oileus,* or Oileus giant owl, rests in a South American rainforest.

Apollo
This butterfly, also known as the mountain Apollo, typically lives in colder, alpine regions of Europe and central Asia. Its hairy white body prefers hills, mountains and flowery alpine meadows, and is often seen visiting thistles.

ALL PHOTOGRAPHS:
Blue tiger
The species *Tirumala limniace* is native to South and Southeast Asia,
preferring scrub forests, dry deciduous woodlands and humid subtropical
forests. It migrates during the monsoons in southern India – the majority
of individuals in the migration are males.

Banded orange heliconian
This orange tiger spreads from Mexico to Brazil and Bolivia, but is occasionally spotted in Kansas and Texas, too. It prefers open spaces in forest clearings, pastures and riverbanks.

ALL PHOTOGRAPHS:

Arctic fritillary

The species *Boloria chariclea*
is found in the colder, northern
parts of the European and North
American continents, often in
alpine meadows, streams and bogs.
In the summer, the males patrol
the valleys and edges of bogs,
looking out for females. When
spotted, they approach them.

Rajah Brooke's birdwing
This magnificent insect is Malaysia's national butterfly. A native to the tropical rainforests of several countries in Southeast Asia, the Rajah Brooke's birdwing prefers sandy banks of rivers and hot springs. These males are gathered in a group drinking mineralized water from puddles.

Blue morpho
The iridescent blue butterfly, *Morpho peleides*, lives in the tropical rainforests of Central and South America. It can usually be found resting on the forest floor, and low trees or shrubs, except for when searching for a mate, when it will fly as high as the treetops and beyond.

LEFT:

Comma

Polygonia c-album, or the comma, is a woodland butterfly that lives across Europe, northern Africa and Asia. However, in the summer, it is often seen in gardens and other habitats searching for nectar and rotting fruit before hibernation.

ABOVE TOP:

Silver-washed fritillary

This fritillary is commonly found in woodlands in Europe, northern Africa and across Asia, including Japan. Often, it can also be spotted above trees and in the open countryside.

ABOVE BOTTOM:

Map butterfly

This spring-form butterfly, which looks like a fritillary, lives in small, partly rural open fields that are surrounded by trees. It is spread across Europe, Central Asia and Russia, up to the fringes of Japan.

Common blue
This species of small blue butterflies is found in Europe, northern Africa and Iran, among grassland habitats and shrubs. It has also been introduced to southeastern Canada. Common blue populations have decreased mainly due to habitat loss.

Dead leaf
The *Kallima inachus* species, commonly known as the orange oakleaf or dead leaf butterfly thanks to its dry autumn leaf appearance when closing its wings, lives across Tropical Asia from India to Japan. Not only does it have the same colour as a dead leaf, it has the same shape and dark veins too, giving it the perfect camouflage.

Common buckeye

This species, *Junonia coenia*, lives in warm, open habitats including forest clearings in the southern United States, Mexico, Cuba and Bermuda. However, common buckeyes migrate north in late spring and the summer, and can also be temporarily found across the United States and into southern Canada.

LEFT:

Painted lady

These colourful butterflies are found on almost every continent except South America and Antarctica. Painted ladies are resident in warmer regions but migrate in spring and sometimes autumn. The widespread migrants usually prefer open, sunny areas like fields and gardens.

ABOVE:

Viceroy

This North American species is found in meadows and humid woodlands, marshes and swamps. It is common in the United States and some parts of Canada and Mexico. Kentucky named the viceroy as its state butterfly in 1990.

ABOVE:
Richmond birdwing
The Australian Richmond birdwing lives in subtropical rainforests near or in the coastal areas of southeast Queensland and northern New South Wales. These birdwings are found along the edges of rainforests, especially close to the flowering plant of the genus *Lantana*.

OPPOSITE:
Malachite
This tropical butterfly ranges from the open subtropical forests in northern South America and Central America to the mango, citrus and avocado orchards in the southern United States. It has been seen flying to flowers as high as 12m (38ft) in the canopy.

Ulysses
This striking blue insect lives in suburban gardens and tropical rainforests in and around Australia, including New Guinea and the Solomon Islands. The Ulysses butterfly is a tourism emblem for Queensland in Australia.

Silver-studded blue
This small blue butterfly can be found in mosslands, heathlands and limestone grasslands across Europe and Asia, from the foothills of the Himalayas to the tips of the Sahara in Africa. In natural light, blue butterflies look similar to our eyes, but different species living in the same habitat can detect some ultraviolet light that helps them recognize their own species.

Spanish moon moth
The moth species *Graellsia isabellae* is native to Spain and France, living in pine forests. A population of the Spanish moon moth also exists in Switzerland, where it was introduced.

ABOVE:
Emerald swallowtail
This banded emerald green
tropical species is native to the
primary forests of southeast Asia.

OPPOSITE TOP:
Ceylon rose
Also known as the Sri Lankan
rose, this swallowtail butterfly
is native to the rainforest in
southwestern Sri Lanka.
However, its natural habitat is
constantly being destroyed by
human activities and it is
critically endangered.

OPPOSITE BOTTOM:
Peacock
A common garden visitor, the
peacock butterfly can also be seen
in parks, meadows and woods
across Europe and temperate Asia.

Monarchs
The migratory monarch lives from the northern parts of South America all the way up to southern Canada. During spring and summer they glide through meadows, whereas in winter they live in higher altitudes. Monarch butterflies are also found in northern Africa and Spain, and reach as far as Australia, New Zealand and the Pacific Islands.

Spanish festoon
The species *Zerynthia rumina*,
found from northwest Africa to
southeast France, likes dry scrubby
areas, woods and grasslands.

BELOW:
Question mark
Polygonia interrogationis, known as the question mark, is native to North America. It lives in wooded areas and city parks, usually in moist areas. When folding its wings, the question mark stays well-camouflaged by resembling a dead leaf.

RIGHT:
Purple emperor
This magnificent butterfly spends most of its life high in the tree tops of woodlands from southern Britain to the Himalayas, and from the Sahara to central China. Females only come down to the ground to lay eggs, while males sometimes drink from puddles.

Zebra swallowtail
This North American species lives up to six months in a moist woodland habitat near rivers, swamps and brushy fields. It can be found across eastern United States and southeast Canada.

ABOVE:
Forest giant owl
This cryptic owl butterfly instantly blends in with its rainforest and forest habitat from Mexico to the Amazon River in South America. It is also seen on banana plantations, where it is regarded as a pest.

RIGHT:
Kaiser-i-Hind
Teinopalpus imperialis, or Kaiser-i-Hind, lives at high altitudes, 1,800–3,000m (6,000–10,000ft), in wooded areas across the eastern Himalayas to northern Vietnam and the Chinese Sichuan province.

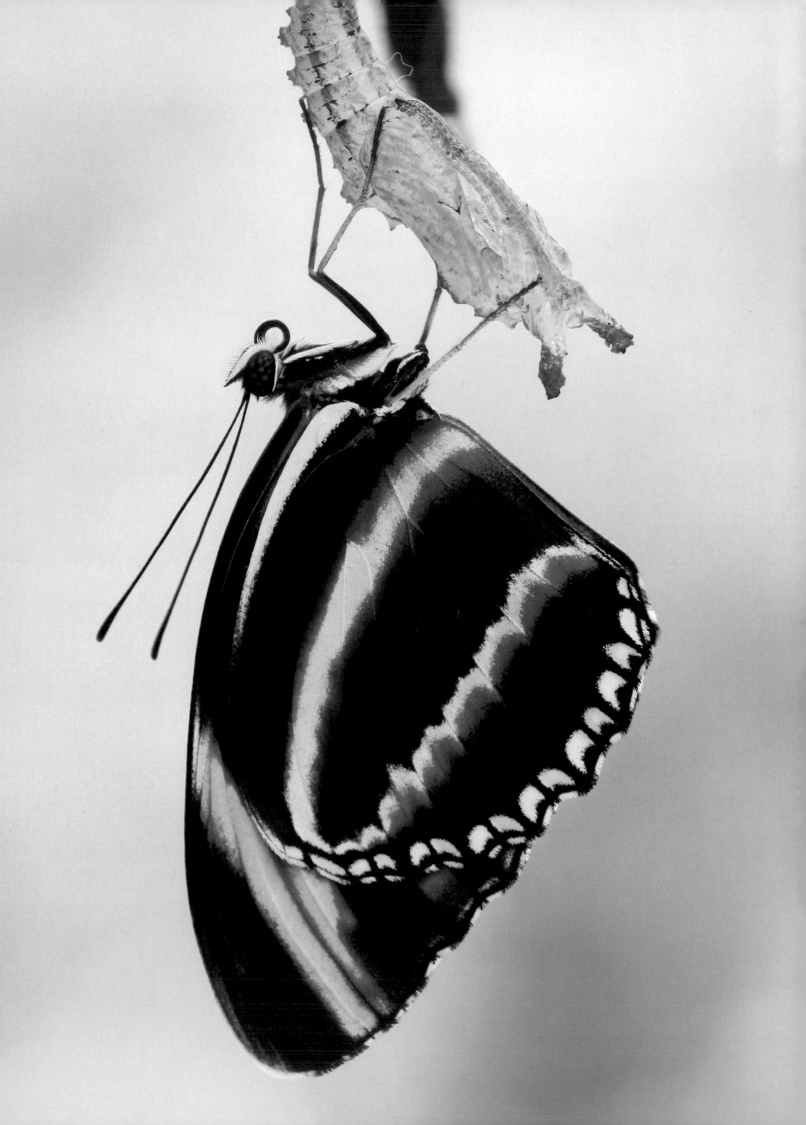

Life Cycle

Most butterflies and moths go through a four-stage process known as metamorphosis: changing from egg to larva to pupa and finally to imago, or adult. Depending on the species, these insects can lay hundreds of eggs – each about 1–3mm (0.04–0.12in) in diameter – which hatch into larvae, commonly known as caterpillars. Caterpillars don't do much but munch on their favourite plants with their chewing mouthparts. As they eat, they grow by molting their outer skin, or exoskeleton, up to five times. Each stage of development in between every molt, or shed, is called an 'instar' until they become pupae. At the pupal stage, known as chrysalis in butterflies, these insects are immobile and appear to be resting. In contrast, moth caterpillars typically spin cocoons from silk around themselves, while others dig holes in the ground. This stage is where the pupa turns into a mushy soup before emerging as a completely transformed adult, with the often colourful, scaly wings.

The males generally emerge first and tend to be smaller and more colourful than the females. On average, most adults will live only two to four weeks, while a few species may live as long as a year. As adults, both are ready to find a mate, reproduce and start the life cycle again.

OPPOSITE:
New life
A newly emerged banded orange heliconian clings onto its empty pupal case.

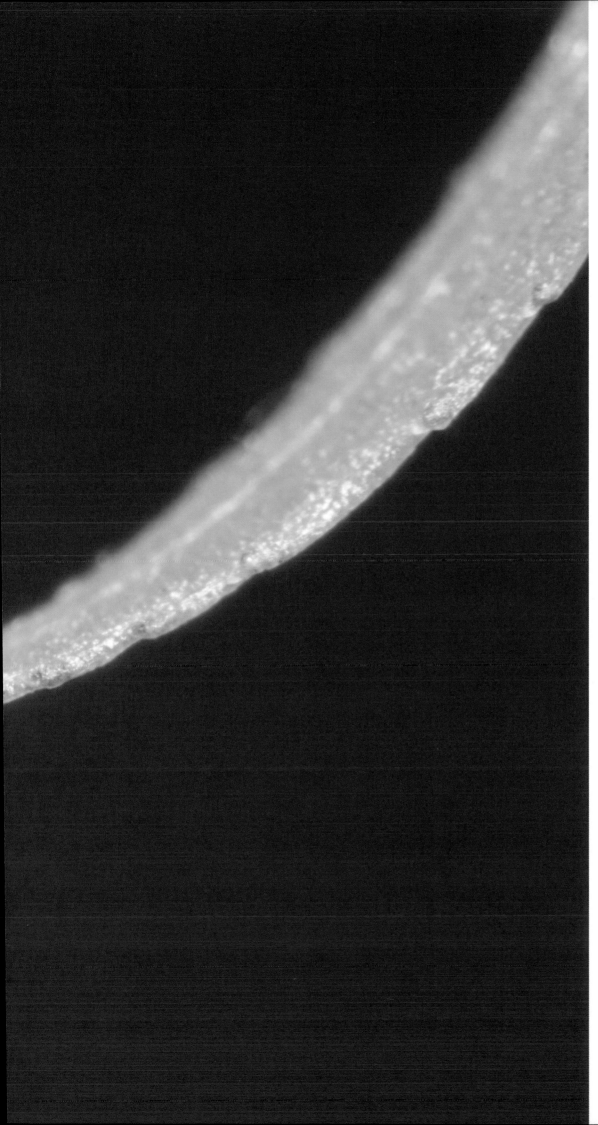

Life begins
The viceroy butterfly begins its life as a tiny round, pale-green egg attached to the tip of a willow leaf. Every butterfly species lays distinct eggs that vary in size, texture, shape and colour. This egg looks similar to growths that occur on the leaves of willows.

ABOVE:
Attachment
Butterfly eggs are normally attached to a leaf with a special sticky fluid made by females. On top of each egg, there are tiny funnel-shaped holes called 'microphiles', where water and air enter. Many eggs start out light-coloured but become darker before the caterpillar hatches out after 3–7 days. This white, round egg, ready to hatch, is that of a glasswing butterfly.

OPPOSITE:
Chrysalis
The Greek word for 'gold' is *chrysós*, which lends itself to the specific name, *chrysallís*, for the metallic-gold colour found in the pupae of many butterflies.

OVERLEAF:
Larval form
A caterpillar is the larval stage of a moth or butterfly. As caterpillars grow, their worm-like bodies change. These Apollo butterfly caterpillars start off as black and are well-camouflaged, and then develop two rows of orange dots as they mature. Fully grown, they pupate in the ground with a loose cocoon for two months.

OPPOSITE:

Protective casing

A chrysalis of a peacock butterfly may be grey, brown or green. Its dents all over help it stay camouflaged by looking like a leaf. Within this protective casing, the caterpillar's body is radically transformed, eventually emerging as a butterfly.

ABOVE:

Sit and wait

A common rose butterfly perched on a branch near its empty chrysalis. Before butterflies can fly away, they must sit and wait for their wings to expand and harden.

Simple eyes

All caterpillars have 12 simple eyes, or ocelli, arranged in a semicircle on the head. These organs only detect changes in light level. In contrast, butterflies have large compound eyes made up of thousands of individual tube-like lenses called ommatidia. Each one of these sends an image signal to the butterfly's brain, which stitches together the information up into a single picture.

Prolegs

This Rajah Brooke's birdwing
caterpillar, like any other insect,
has six true legs located behind its
head. The rest, two to five pairs,
are false legs known as prolegs.
These help caterpillars cling, move
and climb onto plants, and are
lost when they transform into a
butterfly.

Pupal stage

A Rajah Brooke's birdwing
caterpillar enters its pupal stage.
Once fully grown, this caterpillar
turns greyish brown with pale
spikes. Along the back and two
sides, its body is adorned with
yellowish-beige and grey tubercles.
Members of the Papilionidae
family also have an organ – shaped
like the forked tongue of a snake
– to deter predators, which can be
turned inside out behind the head.

Drying off
Rows of butterfly chrysalides – some empty and others intact. When a butterfly first emerges from its chrysalis, its wings are small and wet. This means it must first dry its wings near the empty pupal case before it can fly.

Hair-like setae
Caterpillars, butterflies and moths are often described as hairy or spiky. But only mammals like us have actual hair – an insect's tiny hair-like structures are called setae. These are protrusions of chitin, the same material from which the exoskeleton is made. Caterpillars, like this spiky grey cracker, are able to sense touch through their setae and tentacles.

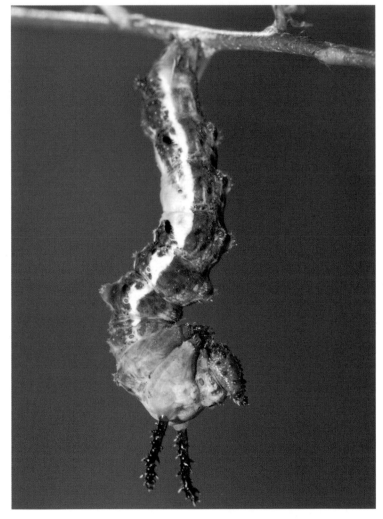

Molting into a chrysalis
This zebra swallowtail caterpillar is eating a pawpaw leaf – its favourite food plant. When the caterpillar is fully grown, it attaches itself by a thin girdle of silk before it molts into a brown or green leaf-like chrysalis. The chrysalis has breathing holes called spiracles.

Bird dropping
A mature viceroy caterpillar has found a suitable place to pupate by hanging upside down like a letter 'J'. To camouflage itself as a caterpillar and a chrysalis, it mimics a bird dropping.

Milkweed muncher
Monarch caterpillars start off as pale green and gradually become white, yellow and black. They munch on milkweed – their favourite plant – which contains chemicals poisonous to birds and other predators. These chemicals make both the caterpillars and butterflies distasteful and poisonous. Monarch caterpillars can grow up to 5cm (2in) long.

Spiky caterpillar
Map caterpillars are black with yellow spots and orange spikes on their backs.

Transformation
A map chrysalis is pale brown marked with olive, hung with the help of a silk button made by the caterpillar. As the chrysalis becomes transparent, the beautiful butterfly can be seen inside. Once it is completely transparent, the butterfly finally emerges from its chrysalis.

Strings of eggs
Female butterflies curl their abdomens towards a leaf or plant to gently lay their eggs on them. This spring-form map butterfly is laying her pale green eggs in long strings, underneath a stinging nettle leaf – the caterpillar's food plant. These strings of eggs likely mimic nettle flowers.

Temperature changes
These red admiral caterpillars, feeding on a stinging nettle, are black with white spots and spikes. However, research has found that temperature affects the colour of these caterpillars, and their colouration can vary. Red admiral caterpillars are about 2.5cm (1in) long.

Vine trickery
Australian Richmond birdwing caterpillars grow up to about 7cm (2.8in) by crunching on two native *Pararistolochia* vine species. As they grow, they travel far to avoid being eaten by their siblings, and their colour changes from black to pale grey, or shades of black. Often these caterpillars munch on the invasive Dutchman's pipe – a deadly vine that tricks the butterflies into laying their eggs on it and kills caterpillars.

Banana leaf
Young forest giant owl caterpillars on the underside of a banana leaf – one of their food plants. The female lays a large number of eggs and these small caterpillars hatch after three to five days. Their appearance changes remarkably as they grow.

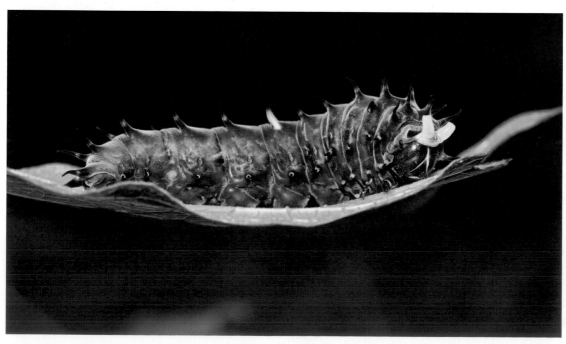

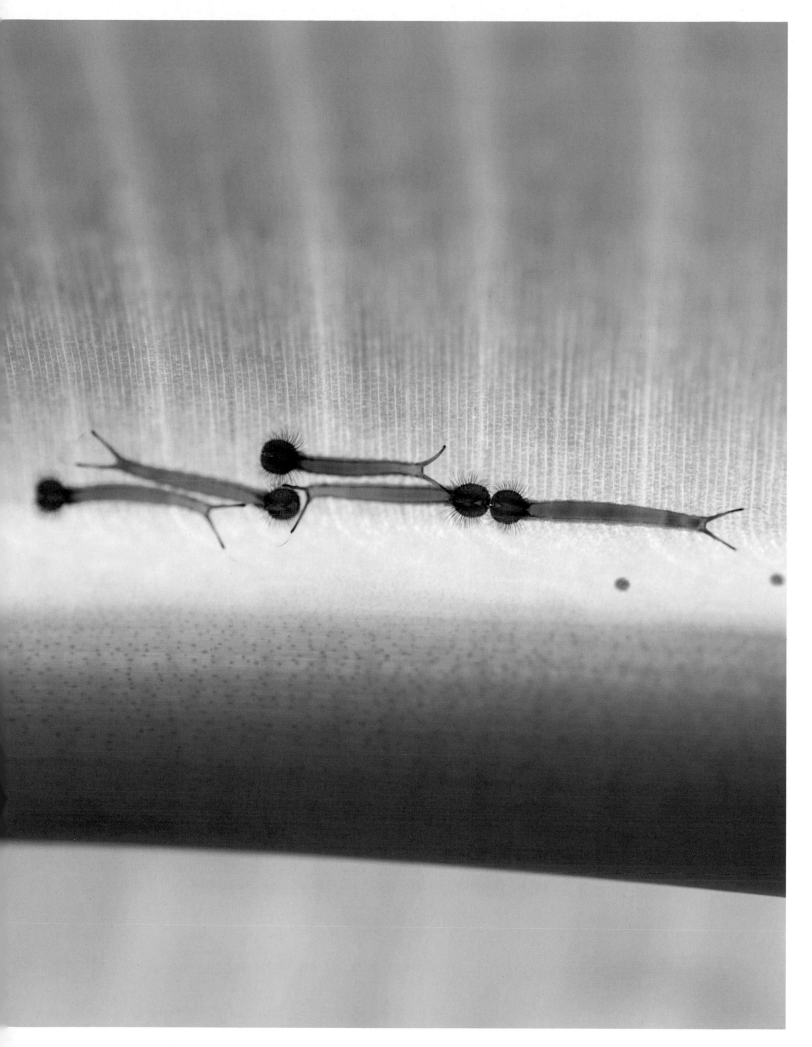

RIGHT:
Short life
A blue morpho emerges from its pupa case – its wings are small and wet, meaning that the butterfly cannot fly yet. Its entire life cycle lasts 115 days and it only lives for about two weeks as an adult.

FAR RIGHT:
Pumping out
A newly-emerged forest giant owl butterfly is pumping out its wings on the empty pupal case. The butterfly hangs upside down for haemolymph – the insect equivalent of blood – to flow into the wings. When ready, the butterfly can fly off in search of liquid fuel.

Behaviour

You'll normally find butterflies on flowers, bathing in the sun, and flitting about. As soon as these insects become adults, they are constantly on the lookout for a mate. Many male butterflies have adapted behaviours to increase their chances of finding a mate, or even deterring others from mating. Some patrol hills, or display in the air, on the ground or a perch, using a distinctive set of odours by flapping their often colourful and attractive wings.

As adults, both butterflies and moths lose their chewing mouthparts and are left with a long tube-like mouthpart that can only slurp up liquids. These insects are specialist bloom feeders, visiting flowers to feast on their nectar. Some also feed on pollen. Unlike most day-flying butterflies, moths generally feed at night – though some slurp up nectar in broad daylight and others go completely without eating. While feeding, grains of pollen stick to the legs and bodies of these insects and cling there during their journey from flower to flower. In this way, many plant species are pollinated both during the day and under the cover of darkness. But some species need additional nutrients, especially males for mating, and have learnt to sip these nutrients elsewhere. So except for plants and flowers, you'll also find these beautiful creatures in the most bizarre places.

OPPOSITE:
Forget me not
Only living for about three weeks, common blue butterflies mate immediately after meeting, usually without any courtship behaviour. This couple – a female and the lighter blue male – is mating on a forget-me-not flower.

Liquid fuel
This bright orange eastern common butterfly is rarely seen visiting flowers. Instead, along with other insects, it is slurping up a rotting persimmon fruit, which has a sweet, honey-like flavour. As they overwinter as adults, these butterflies need additional nutrients rich in amino acids to survive.

Strictly liquid
A common buckeye drinks nectar from a zinnia flower. As a caterpillar, this species devours plants rich in a bitter-tasting chemical to keep ants, birds, wasps and other predators away from itself, and as a butterfly. Butterflies, however, are restricted to a liquid diet because of their straw-like mouthpart, called a proboscis.

ABOVE TOP:
Mimicry
Native to southeast Asia, the swallowtail species *Papilio memnon*, or great mormon, has at least 20 female forms and imitates other species to avoid predators. This female's slow and erratic flight pattern often makes her look like she is dancing, while looking to feed on flowering bushes or to lay her eggs.

ABOVE BOTTOM:
Warning colours
The plain tiger butterfly, also known as the African queen or *Danaus chrysippus*, often flies slowly and close to the ground so predators can recognise its wings' bright colours. This warns them to stay away. The plain tiger stores toxins in its body from poisonous plants, making this butterfly poisonous and distasteful.

RIGHT:
Mate guarding
A female green-veined white butterfly flying through a meadow in the United Kingdom. When the *Pieris napi* species mates, the male inserts a chemical along with his sperm. The scent of this chemical repels other males from mating with the same female.

LEFT:

Nectaring

A ceylon rose, or Sri Lankan rose, butterfly nectars on a hibiscus flower, while constantly fluttering its wings to maintain its position.

ABOVE:

Pollinators

When butterflies sip nectar, their bodies collect pollen unwittingly. This is carried to other plants, helping them breed and develop new seeds. However, less pollen grains stick on butterflies' bodies than on bees, making bees more efficient at transporting pollen between plants.

Migration
Every year, millions and millions of overwintering monarch butterflies arrive in the west-central state of Mexico, Michoacán, coinciding with the Day of the Dead. Locals believe these stunning butterflies are the spirits of their ancestors coming to visit them.

OPPOSITE:
Scent attraction
Blue tiger males often come together in one place to attract a mate
with their scent. By using their protruded brushes found on the tip of
their abdomens, they stroke the pouches on their hindwings. These
pouches, filled with scent scales, scatter the males' scent everywhere,
seducing females.

ABOVE:
Resting alongside
When searching for a mate, the blue morpho will fly everywhere in the
forest – even over the tree tops. Once a male finds his mate, he will chase
the female around a small area, usually flying in a circular pattern.
Mating can last from eight hours to three days, so these butterflies often
rest on plants alongside each other.

ABOVE TOP:
Mud-puddling
Males of the yellow and sulphur species, including pale and apricot sulphurs, are commonly found in large groups sipping moisture from wet sand. This activity is known as mud-puddling and is mostly seen in males, depending on the species. The behaviour occurs before mating season and also when the weather is hot.

ABOVE BOTTOM:
Puddling
This African butterfly, Judith's striped forester, 'puddles' in a rainforest in Ghana.

RIGHT:
Nuptial gift
Before mating season, some males from different species 'puddle' together on patches of wet soil that contain salts and other dissolved minerals. Male butterflies need these minerals to produce pheromones to attract mates and to pass them to females during mating as a nuptial gift. This nourishment helps females produce and lay fertile eggs.

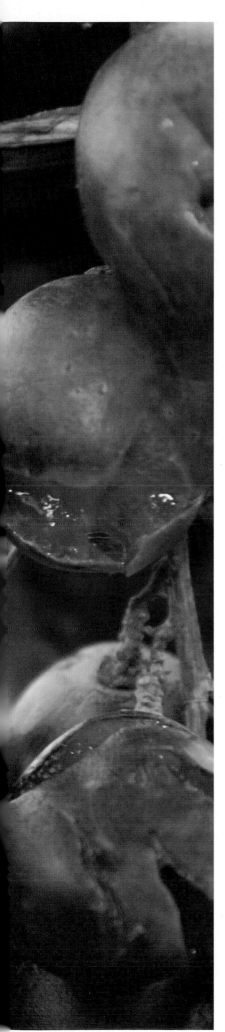

LEFT:
Big appetite
Malachite butterflies have a voracious appetite and often slurp up many liquids – flower nectar, rotting fruits, animal carcasses and bat dung – all day. The butterfly on the left sits on overripe fruit sucking up all the nutrients it needs for flying and reproducing.

BELOW TOP:
Lapping it up
A Rajah Brooke's birdwing uses its proboscis to lap up sweet nectar from these tube-like flowers. Its proboscis looks like a long straw coming out from its head – this helps the butterfly reach the nectar found deep within a flower. When not in use, the proboscis coils up tightly under its head.

BELOW BOTTOM:
Decaying flesh
Glasswing butterflies slurp up nutrients from a dead paper kite butterfly. Dead animals are an excellent source of nutrients for males because they contain salts and amino acids, which are needed for successful mating.

Fruit feeders

Unusually, some butterfly species feed on rotting fruit. This species, the Peleides blue morpho, is never found on flowers slurping nectar. Instead it can be seen sipping the nutrient-rich juices of rotting fruit. When there is no fruit around, it looks for tree sap.

External heat
A common blue male basks
in the early morning sun.
Butterflies must spend time
in the sun to raise their body
temperature with this external
heat source. Besides basking,
some butterflies can also warm
themselves up by 'shivering'.

ABOVE:

Basking

A spring-form map butterfly sunbathes while perched on a plant. This behaviour is known as basking and sees butterflies using their wings' surface to absorb the heat from the sun, giving them energy. While most butterflies bask with their wings spread, some species 'lateral bask' by keeping their wings closed.

LEFT:

Warming up

This peacock butterfly rests on a sunny spot with its wings open. Before butterflies can fly well, these cold-blooded insects have to warm their bodies up to about 30°C (86°F) through direct sunlight. Their wings have veins that allow the warmed haemolymph, or insect 'blood', to be carried around their small bodies.

Dance moves

The Spanish festoon male patrols his territory vigorously until he spots a female. Once she approaches, he follows her and agitatedly flies around her – doing a short dance. He catches her between his folded wings and they fall together onto a plant, where they mate.

Figure-of-eight dance

A banded orange heliconian male will search impatiently for a female. As soon as he spots her, he chases her until she lands on the ground, usually underneath a tree or bush. He flutters around her, with a figure-of-eight dance move, before sitting beside her. If the female is impressed she remains still and the couple mate.

LEFT:

Smell

A zebra swallowtail flies towards a nectar-rich flower. Butterflies are attracted to sweet-smelling flowers through their antennae, or feelers. These sensory organs detect chemicals in the air – scents of flowers or even a potential mate. Antennae can also help butterflies with balance and detecting motion.

ABOVE:

Swing-hoverers

A hummingbird-looking moth hovers over flowers to suck nectar with its long proboscis. While feeding, like their avian namesakes, hummingbird hawk-moths move their wings very quickly and make a humming noise. Their tails open like a fan and can collect nectar, making them good pollinators. They also move from side to side, which is called 'swing-hovering' or 'side-slipping'.

OVERLEAF:

Puddling

Monarch butterflies need moisture on hot days but cannot land directly on water to drink. Instead these butterflies sip moisture and dissolved minerals from the surrounding muddy, wet ground, an example of 'mud-puddling' or 'puddling'.

LEFT:

Roosting

A group of Idas blue butterflies, or northern blues, roost high on plant stems at twilight. At night, and during wet or chilly weather, butterflies become inactive, close their wings and rest. Often in groups, Idas blues specifically rest at the top of grasses with their heads facing downwards to avoid predators such as mice.

ABOVE TOP:

At rest

A ringlet butterfly rests on a fern in southeast England. Overnight or in rainy weather, ringlets find safe places to rest, usually surrounded by many plants and near bushes. They can also be seen sitting on stems or leaves after mating.

ABOVE BOTTOM:

Communal roosting

These upside-down zebra longwings, or zebra heliconian butterflies, gather in the early evening to roost for the night on twigs, 1–2cm (3.3–6.6in) high. The groups range from just a few individuals to up to 30 butterflies and have a social hierarchy, with the oldest allowed to choose the best spots.

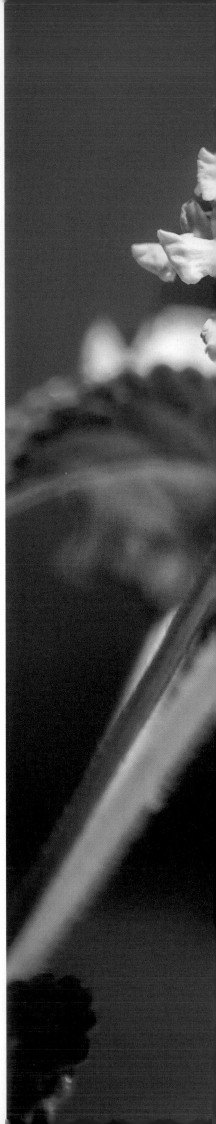

Small tree fan

A Ulysses butterfly lays its round white eggs one by one on a leaf that its caterpillars will eat. These females prefer to lay eggs on small trees, usually no taller than 2m (6.6ft).

RIGHT:

Colour changes

Butterflies do not see the world in the way that we do. Similar to bees, butterflies can detect some ultraviolet light and so flowers appear to have distinct areas that guide the insects directly to the nectar.

LEFT:

Salty snacks

A male purple emperor drinks up salt minerals from a path in the summer. Males spend most of their time in the tree tops, defending their territories from intruders. However, they will occasionally come down to the ground to sip liquids from aphid honeydew, tree sap, animal dung, urine and dead animals.

ABOVE TOP:

Taking a drink

Most often males will 'puddle' together on mud, which contains many nutrients that are unavailable from flowers. When they cannot find mineral salts, both butterflies and moths slurp up salty secretions from animals' skin, eyes and nostrils.

ABOVE BOTTOM:

Slurping minerals

A male zephyr blue slurps up water, minerals and salts from the surface of the mud in the Pontic Alps in Turkey.

LEFT:
Feet-tasting
Butterflies taste plants with their feet – not their proboscises. Their feet detect chemicals on the surfaces they land or walk on. This helps a butterfly sense tasty liquids or identify host plants for their caterpillars.

ABOVE TOP:
Plant selection
A pair of painted ladies slurp up nectar from flowers and aphid honeydew. Research has found that these butterflies usually pick the plants that have the most nectar available to them to lay their eggs on, though this doesn't necessarily mean their brood will have a higher chance of making it into adulthood.

ABOVE BOTTOM:
Overripe fruit
Very fond of overripe fruits, these dead leaf butterflies gather in large numbers to feed from a plate put out for them in China, where they are considered rare.

Lekking
A group of glasswing males will frequently gather together, or form 'leks', to find a mate. This competitive behaviour occurs in shaded corners of the rainforest, where males release pheromones to attract a female.

Picture Credits